Flowers of the Caribbean

G. W. Lennox and S. A. Seddon

MACMILLAN
CARIBBEAN

Macmillan Education
Between Towns Road, Oxford OX4 3PP
A division of Macmillan Publishers Limited
Companies and representatives throughout the world

www.macmillan-caribbean.com

ISBN 0 333 26968 3

Text © G. W. Lennox and S. A. Seddon 1978

First published 1978

The authors and publishers wish to acknowledge the following
photographic sources:
Bermuda Tourist Board p6
Biofotos p46
Anne Bolt p49
Commonwealth Institute p59
J. P. Free p38, 58
Spectrum Colour Library p56
All other photographs courtesy of the authors.
The publishers have made every effort to trace the copyright holders
but if they have inadvertently overlooked any, they will be pleased
to make the necessary arrangements at the first opportunity.

Printed in Malaysia

2005 2004 2003

21 20 19 18 17

Front cover: Hibiscus (*Hibiscus rosa-sinensis*) Michael Bourne
Page iv Bougainvillea (*Bougainvillea spp.*)

Contents

Introduction

Part 1 Herbs and Shrubs

Allamanda *2*
Amaryllis *3*
Angel's Trumpet *4*
Anthurium *5*
Bermuda Lily *6*
Bird of Paradise Flower *7*
Bougainvillea *8*
Canna *9*
Castor Oil *10*
Chenille Plant *11*
Chaconia *12*
Crown-of-Thorns *15*
Cup of Gold *16*
Firecracker *17*
Ginger *18*
Hibiscus *19*
Heliconia *20*

Hibiscus (Coral) *22*
Ixora *23*
Lantana *24*
Morning Glory Bush *25*
Mexican Creeper *26*
Oleander *28*
Orange Trumpet Vine *29*
Passion Flower *30*
Periwinkle *31*
Petrea *32*
Plumbago *33*
Poinsettia *34*
Shrimp Plant *36*
Spider Lily *37*
Thunbergia *38*
Turk's Cap *39*

Part 2 Trees

African Tulip Tree *42*
Bottle Brush Tree *43*
Bauhinia *44*
Cannonball Tree *47*
Cassia (Yellow) *48*
Cassia (Pink) *49*
Cordia *50*
Dwarf Poinciana *51*

Frangipani *52*
Lignum Vitae *55*
Jacaranda *56*
Poui (Yellow) *57*
Poui (Pink) *58*
Poinciana *61*
Powder Puff Tree *62*
Tree Hibiscus *63*

Part 3 Orchids

Introduction

Anyone visiting the Caribbean for the first time cannot fail to be impressed by the richness and luxuriance of the vegetation. Many of the plants encountered produce colourful flowers, and most visitors soon become keen to develop a better acquaintance with some of the more common plants and their exotic blossoms.

In a book of this nature, it is not possible to describe in detail all the species of flowering plants which the tourist or visitor may see during a brief stay. However, it is possible to illustrate and describe some of the more common forms and this book has been designed to be used by individuals with little or no botanical training as an introduction to the identification of the common flowering plants of the Caribbean region. All the photographs included in this book have been taken of plants growing in their natural surroundings in the Caribbean.

One of the most striking features which immediately impresses visitors, particularly those from more temperate parts of the world, is the abundance of flowering trees. When in blossom, these usually appear to be very exotic. The second section of this book is devoted to the most common of these flowering trees.

Orchids are also common in the Caribbean region. However, such is the diversity of this family of flowering plants that their identification is often very difficult. For this reason, the authors have considered the orchids only briefly in the third and final section of the book. Anyone wishing to take the study of the orchids further must necessarily turn to the many detailed books specialising in this particular group of plants.

Hints for non-botanists

Herbs, shrubs and trees
Flowering plants are the most complex and advanced forms of vegetation on the earth's surface. Other less advanced forms of plant life include the algae, fungi, ferns and conifers. Flowering plants can be

grouped, somewhat artificially, into three categories, namely herbaceous forms (herbs), shrubs and trees. The distinction between these groups is not always easy to see. However, the following definitions are given:

Herb This is a plant whose stem is not woody or persistent. After flowering, the stem dies down. The stem is usually green and has a photosynthetic function. Herbs vary in height from a few inches to 'tree-like' forms such as the banana plant.

Shrub This is a perennial plant whose stem is woody and in which new growth is made each year until the maximum height is reached. In any individual shrub, a number of stems arise from near the ground. Generally speaking, shrubs never grow as tall as trees.

Tree This is a perennial plant with a single, self supporting, woody main stem or trunk. Branches usually develop some distance from the ground. Growth takes place each year and vertical growth is accompanied by secondary growth, which causes an increase in diameter of the trunk. The trunk is covered by a protective bark. Sometimes it is difficult to distinguish large shrubs from small trees.

The flower

The flower structure has a reproductive function and, because of this, it is often brightly coloured and sweet smelling, mainly to attract pollinating agents such as insects. Normally, the bright colours are provided by the petals although, in some cases, it is not the petals but other floral structures such as leaves and bracts which are conspicuously coloured. The bright colours of the poinsettia plant arise from the red leaves surrounding the small, insignificant flowers; closer observation is required to see these latter structures. The bright red 'flowers' of the red ginger plant are formed from the coloured bracts. Once again, the real flowers are small and inconspicuous and the observer must look carefully to see them.

The flower bears the reproductive organs and, in many cases, both male and female parts are found in the same flower. The male parts are called **stamens** and the female parts are called the **carpels**.

Some flowers carry either male or female parts only. For example, the maize plant bears separate male and female flowers. In some species, separate male and female plants exist. For example, the paw-paw tree (or papaya) has individual male and female plants and only the female tree bears fruit, after pollination and fertilisation have

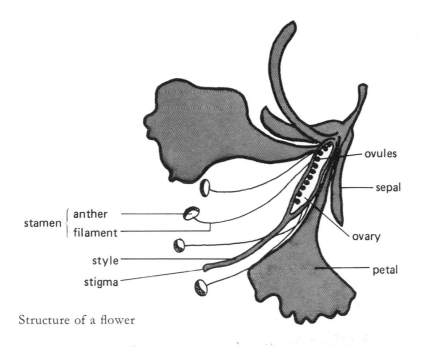

stamen { anther
        { filament

style

stigma

ovules

sepal

ovary

petal

Structure of a flower

successfully taken place. Sometimes, papaya trees bearing herma-
phrodite flowers (both male and female parts in the same flower) do
occur and in some cases the flowers have been known to change sex.

Terminology

All the flowering plants known to science are grouped or classified
in a specific way. It is not the aim of this book to look at the details
of this aspect of plant study called taxonomy. However, for each
plant described in the text which follows, the family name together
with the scientific name (genus and species) is given. These terms are
provided in case the reader wishes to carry out further and more
detailed studies or is tempted, on return to his country of origin, to
obtain particular pot specimens for growth in the home or hot house.

Countries of origin

Many of the more conspicuous flowering plants of the Caribbean
originated in other parts of the world; the region of origin is indi-
cated for each specimen described in the text. As one would expect,
a great many plants now common throughout the Caribbean are
native to Central and South America but many others have been
brought from as far afield as South East Asia and tropical Africa.

1 Herbs and Shrubs

Canna Lilies (*Canna generalis*)

Family   Apocynaceae

Allamanda (*Allamanda cathartica*)
Other names   Buttercup Flower, Golden Trumpet,
Yellow Allamanda, Yellow Bell

Each flower forms a large, golden-yellow, trumpet-shaped structure between four and six inches across. At the base of the internal surface of each petal is a series of orange or brownish lines called honey-guides which run downwards into the centre of the trumpet. The leaves are dark green with a shiny, thick, waxy covering and are pointed at the ends. Allamanda grows as a shrub in most of the main islands of the Caribbean and is particularly common as a specimen shrub in lawns. In some cases, the plant is trained as a vine. A purple variety is also known. All parts are considered to be toxic. It originates from Brazil.

Family   Amaryllidaceae

Amaryllis (*Hippeastrum puniceum*)
Other names   Barbados Lily, Lent Lily

This plant grows from a bulb which gives rise to the long, blade-shaped leaves. It flowers throughout the late winter and spring months. There are a number of different colour forms. Each flower stem attains a height of one or two feet. The individual flowers are trumpet-shaped and each is made up of six petals which fold backwards at the tips. The blooms have the ability to stay fresh for a number of days after being cut and, for this reason, they are often used in indoor floral decorations. In more temperate countries, specimens are grown indoors as pot specimens. The plant originates from South America.

Family   Solanaceae

Angel's Trumpet (*Datura candida*)
Other names   Angel's Tears, Daturas

Each flower forms a white, trumpet-shaped structure which, in large specimens, may attain a length of about ten inches. The 'trumpet' is made up of five joined petals. The flowers hang downwards in large numbers and, towards evening, a shrub in full bloom emits a very strong, heavy scent. The leaves are large, bluish-green in colour and the surface texture is like a fine velvet. Both leaves and flowers are very poisonous. The shrub originates from Central America and flowers at various times throughout the year. It grows to a height of up to fifteen feet.

Family Araceae

Anthurium (*Anthurium andraeanum*)
Other names Flamingo Flower, Heart Flower

This is a particularly attractive and exotic-looking flowering plant which grows to a height of about two feet. It is often used as a decorative plant in hotel gardens and cut specimens are used in floral arrangements. Each flower consists of a large, red, heart-shaped, waxy bract called a spathe. This is often between six and nine inches long, from the centre of which protrudes a long, cylindrical structure called a spadix. The spadix is often whitish or pink in colour and the reproductive flowers are arranged along its length. The leaves are large and heart-shaped like the spathe. Numerous colour variations of the flowers exist, ranging from white and pink to orange and deep red. The common colour seen throughout the Caribbean is the red form. The plant originates from Central and South America.

Family Liliaceae

Bermuda Lily (*Lilium longiflorum*)
Other names Easter Lily

These plants produce large, funnel-shaped blossoms which are white in colour. The leaves are dark green but, in soils deficient in iron, they may become yellow. The plants prefer an open and sunny location in the garden and, during the drier months of the year, it is essential to provide them with plenty of water. This lily came originally from Japan and was introduced into Bermuda in 1853. It quickly adapted to conditions on the island and a flourishing trade developed supplying bulbs to America and Europe. Despite a virus disease which decimated stocks about the turn of the century, the trade has continued and, although not as good as before, there is still a great demand from tourists and local residents. An export trade also still exists.

Family Strelitziaceae

Bird of Paradise Flower (*Strelitzia reginae*)
Other names Crane Flower, Queen's Bird of Paradise Flower

This is one of the most exotic flowering plants found in the Caribbean but it is not common everywhere. The flower stalk bends almost at right angles at the top and from this develops a bluish-grey basal sheath. From this sheath the flowers arise. Each flower has three pointed petaloid sepals usually reddish-orange in colour. The male reproductive parts are arrow-head-shaped and blue in colour. The leaves are large and oval-shaped and there is a reddish-orange midrib down the centre of each. This plant originates from southern Africa and grows to a height of four or five feet.

Family Nyctaginaceae

Bougainvillea (*Bougainvillea* spp)
Other names Paper Flower

This is one of the most colourful and common ornamental vines found in the Caribbean. The main colours vary from red, through purple to deep magenta. There are a number of different forms and additional colours such as white, rose, orange and pink are also seen. The true flowers are small structures and rather insignificant, surrounded by the enlarged and colourful bracts. The leaves are almost triangular in shape and the margin has a wavy appearance. The stem is covered in thorns which help the vine to attach as it climbs. In many cases, the plant has been used to cover walls and fences and, in addition, it can also be seen climbing up tree trunks. It is found throughout the Caribbean. The plant originates from Brazil.

Family Cannaceae

Canna (*Canna generalis*)

There are about sixty species in this fairly small family of plants which originates from South America. Horticulturalists have hybridized them extensively and many of the specimens used in garden displays are of no single specific origin. They are found throughout the Caribbean, and are often planted to form large beds of colour. The colour of the flowers varies and both yellow and red forms exist, but probably the most common variety is red, as shown in the photograph. Each plant grows to about five feet in height. The leaves are long and pointed with parallel venation. A variety exists with reddish-purple leaves.

Castor Oil (*Ricinus communis*)

This shrub has many varieties. It is quick-growing and rapidly attains a height of between twelve and fifteen feet. The plant is attractive, not because of its floral structures, but rather as a result of its colourful fruiting bodies and large, palmate-shaped leaves. The fruits are shown in this photograph. They are spherical, prickly bodies and are reddish-brown in colour when mature. The seeds contain ricin, one of the most poisonous substances known. However, ricin is not soluble in oil and, therefore, the plant is of commercial value because of its oil products. It originates from tropical Africa.

Family Euphorbiaceae

Chenille Plant (*Acalypha hispida*)
Other names Monkey Tail, Red Hot Cat Tail

This shrub originates from South East Asia but it is now common throughout the Caribbean. It is easily recognised by the long, pendulous, red flowers which hang down like tails. Each tail is made up of many staminate flowers and the petals are absent. The leaves are pointed and heavily-veined. Shrubs may reach ten feet in height, although specimens seen in hotel grounds and private gardens are usually smaller than this.

Family Rubiaceae

Chaconia (*Warszewiczia coccinea*)
Other names Water Well Tree, Wild Poinsettia

This is the National Flower of Trinidad. The shrub grows to about eight feet in height and its foliage is dark green. When in flower, each shrub is a colourful mass of scarlet-red sepals. This plant grows wild in the forested region of Trinidad. It needs plenty of water and tends to flower during the wet parts of the year. Fairly recently, a double variety of Chaconia with a greater number of bright red sepals was discovered in Trinidad and this has now been successfully cultivated and propagated. The specimen illustrated was photographed on the St. Augustine Campus, Trinidad.

Family Euphorbiaceae

Crown-of-Thorns (*Euphorbia splendens*)

This plant originates from Madagascar (formerly Malagasy) but it is now widely distributed throughout the Caribbean. It grows as a prickly shrub to a height of about three feet and is often used in hedging and as a rockery plant. It flourishes in full sunlight. The flowers are borne on slender stems which themselves grow from spiny, woody branches. Each stem bears between two and four bright red flowers in a small cluster. The plant stem contains a milky sap which is highly toxic.

Family Solanaceae

Cup of Gold (*Solandra nitida*)
Other names Chalice Vine, Golden Chalice, Gold Cup

This plant is a member of the potato family and is a native of Mexico and Tropical America. It grows as a vine and is often used to adorn walls in hotel grounds and private gardens. Each flower forms a large, yellow, trumpet-shaped bloom, the diameter of which may extend to eight inches. The trumpet is made up of five joined petals and brown honeyguides can be clearly seen on the internal surface of each corolla. The flower changes colour while in bloom; when the bud first opens the petals are bright yellow, but they become more reddish as time goes by. Each flower lasts between three and four days. The leaves are dark green with a waxy, shiny surface.

Family   Scrophulariaceae

Firecracker (*Russelia juncea*)
Other names   Fountain Plant

This shrub originates from tropical America but is now fairly widespread throughout the Caribbean. It grows to a height of between three and four feet and, because of its unusual appearance, it is used as a border plant by many gardeners. The leaves are slender and the stem is usually green and photosynthetic. The small flowers appear as red, tubular-shaped structures, each being about half an inch in length.

17

Family Zingerberaceae

Ginger (*Alpinia purpurata*)
Other names Ostrich Plume Ginger, Red Ginger

This plant is common throughout the Caribbean. It is used in ornamental gardening and in addition to its brightly coloured floral parts, it is also noted for its attractive foliage. The red 'flowers' are made up of waxy looking bracts; the real flowers are hidden inside. They are small, white and inconspicuous. The large, light-green leaves arise on either side of a central stem. The plant blooms throughout the year. It originates from South East Asia.

Family Malvaceae

Hibiscus (*Hibiscus* spp)

There are about two hundred species of *Hibiscus* and types belonging to this genus are found in great abundance in gardens throughout the Caribbean. The size and colour of the flowers vary, but common colours are red, white, pink and yellow. The petals of each flower form a large trumpet-shaped corolla as much as four or five inches in diameter at the end. Protruding from the corolla is a stamen tube with female stigmas at the tip. The most common form is *Hibiscus rosa-sinensis* which originates from Hawaii, but has now been introduced to most tropical regions. The flowers of this particular species can be used to clean and shine black shoes. One species (*Hibiscus esculentus*) produces fruits known as okra and these are used widely in cooking. The shrub has alternate simple leaves which are coarsely serrated. It blooms throughout the year and prefers a sunny situation. Each flower lasts for only a day, and at nightfall each corolla drops to the ground. *Hibiscus rosa-sinensis* has been used extensively in producing horticultural varieties, a process known as hybridization.

Heliconia (*Heliconia* spp)
Other names Lobster Claw, Wild Banana

There are about forty species of *Heliconia,* often called 'wild banana' in the Caribbean. The plants are related to the edible banana and the leaves are 'paddle-shaped', a characteristic of the banana family. The brightly-coloured floral parts are really large bracts and they arise alternately from the stem. Sometimes the stem hangs downwards (pendant) and in other types the stem is erect. The bracts open to expose inconspicuous purple flowers inside. Very often, pollination is brought about by hummingbirds. The bracts are long-lasting and are often used in floral decorations. This shrub grows wild in a number of Caribbean Islands and fine specimens are to be seen in Hope Gardens, Jamaica. The pendant form above was photographed there. These plants originate from tropical America.

Family   Malvaceae

Coral Hibiscus (*Hibiscus schizopetalus*)
Other names   Dissected Hibiscus, Fringed Hibiscus

This species of *Hibiscus* produces much daintier flowers than the more common forms of the shrub. Each corolla is made up of divided petals, the whole structure giving a lacy appearance. The colour is usually pink and the long stamen tube with the stigmas at the end hangs down several inches beyond the petals. The coral hibiscus is not as common as the other forms of the shrub, but it can be found on most of the islands where it is grown as a specimen shrub in lawns. The leaves are narrower than those of related forms and they are light green in colour with serrated or toothed-margins.

Family Rubiaceae

Ixora (*Ixora macrothyrsa, Ixora coccinea*)
Other names Flame of the Wood, Jungle Flame, Jungle Geranium

This shrub is common throughout the Caribbean and it flowers for most of the year. Each individual flower has four petals; the flowers are grouped together to form large spherical heads, giving the shrub a very colourful and exotic appearance. The common colour is red, but other colours (including white) are found. Each 'flower ball' may be as much as six inches in diameter. The leaves are dark green and very shiny and make an attractive background for the red flower heads. The plant is grown as an ornamental shrub and sometimes as a hedge. Cut specimens have long lasting qualities when used in floral decorations. This shrub originates from the East Indies.

Family Verbenaceae

Lantana (*Lantana camara*)

This is a very common shrub in many of the Caribbean islands. The visitor will often see the plant growing along roadsides and on waste ground where it may attain a height of four or five feet. There are a number of different colour forms. The individual flowers are small, each being about one tenth of an inch in diameter, but they arise in groups producing dense, flattened collections of blooms. The leaves are dark green, small and each leaf margin is distinctly serrated. The leaf veins are particularly conspicuous. The leaves and berries are toxic. In some gardens, this species is grown as an ornamental shrub, particularly in sunny positions. Because of the large numbers of flattened flower heads, butterflies are readily attracted to this shrub. The plant originates from the West Indies.

Family   Convolvulaceae

Morning Glory Bush (*Ipomoea fistulosa*)
Other names   Potato Bush

This shrub is a close relative of the more abundant Morning Glory Vine. It grows to a height of about five feet and is sometimes used as a specimen shrub in lawns. The leaves are simple, pointed and heart-shaped. Although it originates from tropical America, it is now fairly widely distributed throughout the tropics. It produces large, light mauve, trumpet-shaped flowers. The texture of the flowers is often very flimsy and delicate and newly opened blossoms are easily torn by sudden gusts of wind. The flowers soon develop a rather ragged and untidy look, particularly at the ends where they form the edge of the trumpet.

Family Polygonaceae

Mexican Creeper (*Antigonon leptopus*)
Other names Chain of Love, Coral Vine

This plant is a common climber on hedges and fences throughout the Caribbean. The leaves often appear rather a dull green in colour and are heart-shaped with wavy edges. The small flowers grow in great profusion and are reddish-pink in colour. A closer inspection of the flowers reveals that the petals are poorly developed and the bright colours are due to the five sepals present in each flower. The plant originates from Mexico.

Family Apocynaceae

Oleander (*Nerium oleander*)
Other names Rose Bay

This shrub grows up to twenty-five feet tall. However, it is cultivated in gardens and, by careful pruning, it can be kept low and
bush-like. Some gardeners also grow it as a hedge. The leaves are
thin, pointed at the ends and appear dull green in colour. Flowering
is profuse and continuous. The flower colour is usually pink but red,
white and cream flowering shrubs are also seen. This plant originates
from the Mediterranean region and Asia Minor and was well known
to the ancient Romans. All parts of the plant are poisonous and
deaths have resulted from food cooked on oleander wood fires.

Family Bignoniaceae

Orange Trumpet Vine (*Pyrostegia venusta*)
Other names Flame Vine

When in flower, this vine has a very exotic and colourful appearance. It flowers from January to April and can be seen in some of the major islands including Trinidad. It often grows on fences and walls, and extensive growths sometimes cover the roofs of buildings. The flowers are long, thin and tubular-shaped, the orange corolla comprising five lobes which curl back at the ends. The leaves, which have pointed ends, are dark green and tend to curve inwards from the midrib. The country of origin is Brazil.

29

Family Passifloraceae

Passion Flower (*Passiflora* spp)

There are between 350 and 400 species of passion flower plant. They originate from South America but they have been taken to many other parts of the world including more temperate zones. These plants grow as vines, clinging to walls and fences by means of long, coiled tendrils. Although there are many species, the arrangement of the main floral parts is the same. Each flower has five sepals, five petals, five anthers and three stigmas. The flower is considered to symbolise the crucifixion. The colour of the perianth parts varies and both white and scarlet forms are found. Some species produce ball-shaped, edible fruits which are crushed to make a refreshing drink. In this connection, one species in particular (*Passiflora edulis*), is grown commercially for its fruits.

Family Apocynaceae

Periwinkle (*Catharanthus roseus*)
Other names Ramgoat Rose

This is a common flowering plant found throughout the Caribbean. It is often used as a border species in ornamental gardens and flowers continuously. The corolla is made up of five petals and, at the centre where they meet, there is usually a darker, circular pattern. The colours of the flowers vary from pinkish-red to white. The leaves are dark green and have a shiny, waxy surface. The plant originates from Madagascar.

Family Verbenaceae

Petrea (*Petrea volubilis*)
Other names Blue Petrea, Bluebird Vine, Purple Wreath Vine

This vine grows widely throughout the Caribbean. The small, blue flowers grow in sprays and, when in full flower, the green vegetation seems to be covered with inflorescences hanging down in large numbers. The leaves are dull green, each with a prominent mid rib and side veins. When the petals fall, the sepals remain behind and form the 'wings' for the seeds when mature. The plant originates from tropical America.

Family Plumbaginaceae

Plumbago (*Plumbago capensis*)
Other names South African Leadwort

This small shrub grows to a height of two to three feet. It flowers
for most of the year. The flowers are pale blue in colour and they
contrast sharply against the bright green leaves. The corolla is made
up of five petals, each petal is characterised by a central line running
its full length. All parts of this plant are toxic to some extent. The
plant grows best in full sunlight. It originates from southern Africa.

Family Euphorbiaceae

Poinsettia (*Euphorbia pulcherrima*)
Other names Fire Plant, Painted Leaf

This shrub may attain a height of twelve feet. The degree of flowering is dependent upon day length, and because the shrub blooms more effectively when day length is shorter, it is often called a 'short day bloomer'. The bright red structures are leaves; a group of leaves surround the flowers which are small, less conspicuous and yellow in colour. All parts of the shrub are toxic. The country of origin is Mexico, but now this species is common throughout the tropics where it is grown as a decorative garden shrub. In more temperate countries, the plant has become popular as an indoor specimen.

Family Acanthaceae

Shrimp Plant (*Justicia brandegeeana*)

This shrub usually attains a height of between three and four feet. The name is derived from the appearance of the inflorescences which are formed of coloured bracts (usually brown or red) overlapping each other. The flower head, therefore, gives the impression of a scaled or plated structure such as the body of a shrimp. The real flowers are found within the coloured bracts and are tube shaped and white in colour. The plant originates from Mexico and it flowers throughout the year in many Caribbean islands. In many temperate countries, this shrub has become a popular house plant.

Family Amaryllidaceae

Spider Lily (*Hymenocallis caribaea*)

This plant is easily recognised by its unusual flower structure. Each flower has six, thin (usually white) petals and six reddish-brown stamens which are pendulous. The leaves vary in size, usually attaining three or four feet in length. The leaves tend to arise in clumps from the middle of which arises the stem. The head of the stem bears a varying number of flowers. The plant is usually grown as a border specimen and its flowers produce a pleasant scent. It originates from tropical America.

Family Acanthaceae

Thunbergia (*Thunbergia grandiflora*)
Other names Bengal Clockvine, Bengal Trumpet, Sky Flower

This plant is common in many of the islands. It is a vine and is often cultivated in gardens to form an attractive covering for walls and fences. Each flower is formed of five petals, joined to form a trumpet-shaped corolla. The funnel formed by the joined petals is usually yellow in colour and there are distinct honeyguides. The flowers are blue in colour, although a white variety is sometimes seen. Flowering is continuous throughout the year. The leaves are large, heart-shaped and the surface is rough to the touch. The plant originates from India.

Family Malvaceae

Turk's Cap (*Malvaviscus arboreus*)
Other names Pepper Hibiscus, Sleeping Hibiscus

This shrub attains a height of between six and seven feet. The common name is derived from the fact that the flower bears a resemblance to the form of head wear known as a Turk's Fez. Each red flower looks like an unopened bloom of common hibiscus, with the stigma and stamens extending beyond the edge of the petals. The leaves are dark green and serrated at the edge. The shrub flowers throughout the year. The plant originates from Central America.

2 Trees

African Tulip Tree (*Spathodea campanulata*)

Family Bignoniaceae

African Tulip Tree (*Spathodea campanulata*)
Other names   Flame of the Forest, Fountain Tree, Tulipan

This is a large tree which often attains a height of forty or fifty feet. It was first discovered in Ghana (then the Gold Coast) in 1787 and has since been introduced to many parts of the world, including the Caribbean. Many local people believe the tree to have magical properties. The red flowers grow in circular formations, and in the centre of each circle is a collection of crowded buds. Only a few flowers open at any one time. The tree flowers throughout the year. The boat-shaped seed pods are usually large and may attain a length of two feet. Many of the unopened flower buds contain an excess of water, and this can be released when pressure is applied – hence the name of 'Fountain Tree'.

Family Myrtaceae

Bottle Brush Tree (*Callistemom lanceolatus*)

This forms a medium-sized tree belonging to the Myrtle family from Australia. The red blossoms are long and cylindrical in shape and each consists of a large number of flowers arranged along the length of the floral axis. The unusual spiky appearance of each bloom results from the large number of red stamens and this arrangement produces a structure similar to the type of brush used for cleaning babies' bottles – hence the common name of the plant. The leaves are narrow and pointed and tend to be light green in colour. Although the tree originates from Australia, it is now distributed in a number of Caribbean islands including the Bahamas and Bermuda.

Family Leguminosae

Bauhinia (*Bauhinia purpurea*)
Other names Butterfly Tree, Orchid Tree, Ox or Bull Hoof Tree

This is a small to medium-sized tree with characteristic leaves shaped like the hoof of a cow. In some parts of the tropics it is called 'camel's foot'. The flowers resemble orchids in shape, the reproductive parts being exposed and clearly visible. There are a number of different species and distribution is fairly wide throughout many of the Caribbean islands, including Antigua, Trinidad and Jamaica. These trees originate from India and South East Asia but they are now widely distributed throughout the tropics. There is also a yellow-flowering species (*Bauhinia tomentosa*) native to tropical Africa.

Family Lecythidaceae

Cannonball Tree (*Couroupita guianensis*)
Other names Carrion Tree

This is a large tree with a thick, tough bark. The short, crooked branches grow directly out of the main trunk and they support both the large, reddish flowers and the spherical fruits. The tree is named after the shape of the strange fruits which are often six to eight inches in diameter. The flowers are about five inches across and the reproductive parts are completely exposed. The flowers produce a sweet smelling odour which attracts bats and other pollinating agents to feed on the nectar. It has been suggested that the open, saucer-shaped flower is able to reflect echoes back to the flying bats, thus guiding them to the nectar. The fruits, when mature, contain a pulp with an unpleasant odour. Empty fruit cases are sometimes used as calabashes. The tree originates from the northern parts of South America. This specimen was photographed in Hope Gardens, Jamaica.

Family Leguminosae

Cassia (*Cassia fistula*)
Other names Golden Shower Tree, Indian Laburnum,
Pudding Pipe Tree, Shower of Gold

This tree is found throughout most of the Caribbean islands. It reaches a height of thirty or forty feet and is often grown to provide shade. The yellow blossoms are produced in large numbers, each flower comprising five yellow petals. A long, curved pistil extends from the centre of each flower and the stamens are also exposed. After fertilisation, each pistil develops into a long, brown fruit pod which may reach a length of two feet when mature. The leaves are characteristic of this family of plants, being formed of compound leaflets. This tree originates from India where, because of the long fruit pods, it is called the 'Pudding Pipe Tree'.

Family   Leguminosae

Pink Cassia (*Cassia javanica*)
Other names   Apple Blossom Cassia, Pink Shower Tree

This tree grows to a height of thirty feet and, like its related yellow form (*Cassia fistula*) it is often cultivated to provide shade. Each flower has five petals, the colouring of which is pale pink with some white and darker pink markings. From a distance, a tree in full bloom gives a variegated effect. This species originates from Java but it is now widely distributed throughout the tropics, including many parts of the Caribbean.

Family Boraginaceae

Cordia (*Cordia sebestena*)
Other names Anaconda, Geiger Tree, Geranium Tree,
Spanish Cordia

This is a small tree which attains a height of twenty or thirty feet. It is indigenous to the Caribbean and it occurs in most of the islands, where it is often planted to provide shade. It produces colourful orange-red flowers and plum-like, edible fruits. In some of the Caribbean islands it is called Anaconda or Spanish Cordia. The flowers grow in clusters with as many as fifteen blossoms per cluster and the flowers mature first towards the centre of each cluster. Each flower is about one inch in diameter and the petals are paper-like and frilled at the edges. The tree likes dry conditions.

Family Leguminosae

Dwarf Poinciana (*Caesalpinia pulcherrima*)
Other names Peacock Flower, Pride of Barbados

This small tree attains a maximum height of about twelve feet and flowers throughout most of the year. The flowers arise in clusters from the tips of the branches and, although red is the common floral colour, a yellow variety is also found. Each bloom has five petals and the red-coloured flowers have a yellow margin on each petal. The reproductive stamens and pistil are exposed and extend well beyond the corolla. The leaves are 'fern-like', each consisting of a central stem with small leaflets arising on either side; the branches are prickly. The tree is wide-spread throughout the Caribbean. It probably originates from Central America and the West Indies but its exact place of origin has not been decided.

51

Frangipani (*Plumeria rubra, Plumeria acutifolia*)
Other names   Pagoda Tree, Temple Tree

There are many species and varieties of frangipani. The flower colours vary and reddish-orange, pink, yellow and white specimens are not uncommon. This small tree attains a height of about fifteen feet and is distributed widely throughout the Caribbean. It is particularly attractive because of the sweet scent produced by the flowers. The tree tends to flower continuously. Most species shed their dark green leaves during the drier parts of the year. The stem contains a white, milky sap which is very poisonous. The tree develops a characteristic shape since it produces only a short trunk which begins to branch close to the ground. Each new shoot quickly branches into two and thus the tree soon develops a multi-branched appearance. The tree originates from Central America.

Family Zygophyllaceae

Lignum Vitae (*Guaiacum officinale*)
Other names Tree of Life

The blue flower of this tree is the national flower of Jamaica. The mature tree has a rounded, dome-shaped form, the trunk branching a few feet from the ground. The leaves are small and rounded and the blue flowers arise in large clusters at the ends of the branches. The wood is the heaviest of commercial woods and its specific gravity may be as high as 1.25. It is highly resinous and this, together with its density, makes it useful in the manufacture of functional parts such as pulley blocks, cogs and bearings. Gum guaiacum collected from the bark is used in the treatment of arthritis and related conditions and a decoction of the sap wood was used, until more modern drugs were developed, in the treatment of syphilis. The tree is a native of tropical America and although it is particularly common in Jamaica, it is also found in Trinidad and Tobago and Cuba.

Family Bignoniaceae

Jacaranda (*Jacaranda acutifolia*)
Other names Fern Tree

This tree originates from Brazil but it has now been introduced into many tropical and subtropical regions. Trees reach a height of thirty or forty feet and the foliage is characteristically fern-like, the leaves resembling those of the flamboyant. There are more than forty different species of *Jacaranda* but they all have a common floral structure. Each bluish-purple flower is bell-shaped and the blooms arise in clusters. After fertilisation, the fruit bodies are produced. These are flat, hard and dark brown in colour when mature. They release winged-seeds which are dispersed by wind and air currents. This tree is found in various parts of the Caribbean, and in Trinidad and Jamaica it is known locally as the 'Fern Tree'.

Family Bignoniaceae

Yellow Poui *(Tabebuia serratifolia)*

Other names Apamata, Gold Tree

There are several kinds of Yellow Poui Tree and, when in full bloom, they are one of the most striking sights of the Caribbean. They grow to a height of seventy or eighty feet and can be considered among the largest of tropical forest specimens. The wood is extremely dense and, because of its resistance to termites and general decay, it is often used in building and construction work. Frequently, the tree flowers in the absence of leaves and new foliage develops as the blooms drop. When in full foliage, each leaf comprises five to seven leaflets which are silvery-green in colour. Each flower has a trumpet-shaped corolla and the blossoms are grouped together in large numbers. Flowering occurs at the end of the dry season. The tree originates from South America.

Family Bignoniaceae

Pink Poui (*Tabebuia pentaphylla*)
Other names Pink Tecoma, Pink Trumpet, Trumpet Tree

Like its relative the Yellow Poui, this species is one of the tallest forms of forest tree, reaching a height of sixty or seventy feet. The wood is important commercially and is often used in internal building work. Although planted for its floral beauty, it is also grown for supplying shade among coffee and cocoa plantations. The pink flowers arise in clusters of trumpet-shaped blooms and, although this species often attains a great height, it will flower when only three or four years old. It tends to flower in April and May. The tree originates from South America.

Family Leguminosae

Poinciana (*Delonix regia*)
Other names Flamboyant, Flame Tree, Royal Poinciana

This tree is now widely distributed throughout the Caribbean. It often attains a height of forty or fifty feet and, because it grows with a branching habit, it has been used as a shade tree in many of the islands. When in bloom, the flowers arise in dense clusters of reddish-orange colour. Each flower has five petals, one of which is yellowish or white in appearance. After the flowers and feather-like leaves have withered and dropped off, the seed pods remain. They often hang on the tree for several months and mature specimens may attain two feet in length. In some islands, they are used as a fuel for fires. The mature pods are sometimes called 'woman's tongue' because of the rattle the seeds make when the pods are disturbed by the wind. This specimen is often planted along roadsides and this has resulted in the vigorous root systems breaking up the side pavements. The tree originates from Madagascar, although it is reputed that specimens grew wild in Jamaica in 1756. Indeed, it is named after M. de Poinci, a governor of the French West Indies.

61

Powder Puff Tree (*Calliandra inaequilatera*)
Other names Mimosa, Redhead Calliandra

This small tree is characterised by its unusual-shaped flowers. Each flower is bright red in colour and about two inches in diameter. The main floral feature is not the petals but the long, silky stamens. A tree in flower gives the impression that it has large numbers of 'pompoms' scattered among its foliage. The leaves of this tree are 'fern like' and are characteristic of leguminous plants. The leaf shape is clearly illustrated in the photograph. The tree originates from South America.

Family Malvaceae

Tree Hibiscus (*Hibiscus elatus*)
Other names Cuban Bast, Mountain Mahoe

This can be grown as an ornamental shrub or, if not frequently pruned, it will develop into a medium sized tree attaining a height of forty or fifty feet. The leaves are characteristic of the Hibiscus family and are simple and alternate. The flowers are yellow when newly opened early in the day, but they gradually darken to a bronze colour and finally become red before nightfall. They then drop off and new buds open the next morning. If the reader wishes to photograph flower specimens of this species, it should be done before eleven o'clock in the morning when the blossoms are at their best. The plant originates from Hawaii.

3 Orchids

Cattleya Orchid Hybrid

Orchids

As stated in the Introduction, the orchids form a very specialised group of plants and it is not possible, in a book of this nature, to deal with them in any detail. The casual visitor or tourist will see examples of this family of plants in many of the Caribbean islands, but their identification is often a very difficult thing to determine. In some of the botanical gardens, such as Hope Gardens in Jamaica, orchids are displayed with identity labels and it is probably a good idea if the non-botanist first makes an initial study in such places. Numerous books have been written on the Orchidaceae and if the reader wishes to make a serious study it would be better to consult one of the more advanced books dealing specifically with this family.

More than 20,000 species of orchid are now known and during their evolutionary history they have colonised a wide variety of habitats in different parts of the world including both tropical and temperate countries. In many parts of the tropics, they live both on the ground and on the trunks and branches of many forest trees. Such plants are called epiphytes and this habit is known as epiphytic.

The majority of orchids obtain nourishment from the air, rain, or moisture in the soil. They are also able to manufacture sugars and starch as other green plants do and for this they require sunlight. Many epiphytic orchids develop specialised storage structures and some types produce aerial roots whose main function is to absorb moisture.

Although the flower size and shape varies greatly within this family of plants, the general floral arrangement is the same in all species. Normally there are six perianth parts, the outer three of which are usually green and sepal-like while the inner three are petal-like and often very colourful. One of the petals is larger than the other two and is called the labellum. Normally only one stamen produces pollen.

Many orchids have developed a very specialised partnership with insects and, in some cases, a particular species is pollinated only by a single species of insect.

Cattleya (*Labiata*)

One orchid in particular (*Vanilla planifrons*) is grown commercially for its fruits known as vanilla 'beans'. The procedure involved in curing the beans is a complex one and modern methods are similar in some respects to the traditional methods encounted by the Spanish in the sixteenth century when they first made contact with the Aztecs. The biochemistry of curing results in the production of a substance called vanillin and this, more commonly called vanilla, is used as a flavouring ingredient in the manufacture of sweets, cakes, perfume and tobacco. The particular orchid from which vanilla is extracted originates from Central America, but it is now grown extensively in other parts of the world.

The flowers described and illustrated in this book are ones which can be seen in either the grounds of the various hotels or in private gardens viewed in passing. It is hoped that they have given the reader some insight into the variety, richness and luxuriance of the vegetation which can be seen throughout the islands of the Caribbean, the Bahamas and Bermuda.

Visitor's check list

Date	Name of plant	Place where seen	Additional notes

Vistor's check list

Date	Name of plant	Place where seen	Additional notes

Vistor's check list

Date	Name of plant	Place where seen	Additional notes